MIND-BLOWING MATH

MYSTERIES REVEALED

Nansea Cross

MIND-BLOWING MATH
MYSTERIES REVEALED

iUniverse books may be ordered through booksellers or by contacting:

iUniverse
1663 Liberty Drive
Bloomington, IN 47403
www.iuniverse.com
1-800-Authors (1-800-288-4677)

ISBN: 978-1-4917-6213-4 (sc)
ISBN: 978-1-4917-6215-8 (hc)
ISBN: 978-1-4917-6214-1 (e)

Library of Congress Control Number: 2015903306

Print information available on the last page.

iUniverse rev. date: 08/12/2016

CONTENTS

Introduction_____**vii**

Preface_____**xi**
This shows the multiplication table and explains how there is diagonal symmetry in one direction and also explains how my system is going to discover diagonal symmetry in the other direction.

Chapter 1 – The Numbers 0 through 9_____**1**
This chapter starts out by explaining how to reduce numbers and then goes through each number individually, showing the patterns of the multiples and their reductions along with several other interesting patterns. There are numerous charts and graphs to help show all the patterns that are discovered.

Chapter 2 – Recap and Summary of the Numbers 1 through 9____**21**
Chapter 2 shows charts with all multiples and their totals. It also shows patterns with the reductions of all the multiples. It concludes with some astonishing symmetry and patterns. Crazy patterns are revealed by adding the rows diagonally in both directions.

Chapter 3 – Adding Consecutive Numbers_____**45**
Chapter 3 shows an easy way to add consecutive numbers and then goes on to show patterns that are formed by adding consecutive or triangular numbers. There is also a section on the 3, 6, and 9 team.

Chapter 4 – Perfect Squares_____**49**
This chapter shows patterns that are found in the reductions of perfect squares and perfect cubes.

Chapter 5 – Exponents_____53

Chapter 5 takes each number individually and discovers patterns with the individual numbers and their exponents. Negative exponents also reveal patterns in most instances.

Chapter 6 – The Hundreds Board_____63

The chapter on the hundreds board reduces all the numbers and once again reveals several interesting patterns, including adding the numbers horizontally, vertically, and diagonally.

Conclusions_____73

This basically asks the reader, how could all these coincidences exist in all these numbers? Could this possibly be all random or is there some explanation? Personally I like to keep an air of mystery here to bring some fun and excitement into the process of learning and teaching the fascinating subject of math.

Worksheets_____75

INTRODUCTION

As math was a subject that came easily to me, I thought it would be a subject I could teach successfully, but it didn't take me long to realize that most kids don't like math and really struggle with the subject. It was really more like a case of pulling teeth. I somehow mostly ended up with the struggling students who hated math. Despite that, I've managed to spend about fifteen years teaching and tutoring math students. Since so many of my students really just didn't like math, I tried to find ways to make it more fun and interesting. I also always tried to find the easy ways to solve problems. Through this process, I discovered a whole hidden dimension of patterns and sequences in numbers, which adds a whole new element to math. I used this information to try to encourage, motivate, and inspire my math students, many of whom are math phobic or math haters. It's truly mind-blowing how so many patterns exist in math. My goal in writing this book is to get kids more interested in math and to make math fun. For work sheets and lesson plans, you can go to my website: http://www.mindblowingmath.com. Please contact me with any new patterns you find that I have not discovered. Most of the patterns in this book just involve simple addition and multiplication. We will be doing a little subtracting and dividing in the process, as well. For the most part, the work involved is at about a fourth-grade level.

Math is one of the most important subjects a student has to learn, yet so many kids these days really struggle with math. They don't like it, they don't do well with it, and they avoid it at all costs. Unfortunately, the handy calculator has become a crutch to so many students that they don't feel they need to learn math. The knowledge of math is a skill that is definitely needed to survive and thrive in this world.

I was fortunate growing up in that math always came easily to me, and I even enjoyed it. Even today, I love to solve difficult math problems; I guess I'm a nerd in that. Solving puzzles was also something in which I excelled and enjoyed. One other trait that I've been accused of is being very observant. I believe that these three skills enabled me to make the discoveries that I'm about to unveil in this book.

I taught high school math, mainly in a dropout-prevention program, for several years. My students really struggled with the math. Many of them tested out with the skill level of a third-, second-, or even first-grade-level student, but they had to pass high school algebra in order to graduate. That was a very challenging task for me. As my students were enrolled in my classes all through the school year, they were all on different levels and different pages in their books. I even had one class with eight different math courses going at the same time. That definitely wasn't an easy job. I was basically tutoring thirty different students or so per class. Needless to say, I would wrack my brain trying different methods to teach my students subjects that were really way above their skill level.

One huge mistake the educational system is making in the United States revolves around the testing system that's been developed over the past decade or so. Instead of focusing on a few topics per year in which to gain mastery, such as multiplying and dividing, students have to learn several skills in a single school year. Many of these skills are beyond their current learning ability. For example, students in fourth grade are working on the beginnings of algebra and geometry. That's like learning to dance before you are adept at walking. It just does not serve the best needs of our students.

While teaching math to these students, I discovered that there are lots of hidden patterns in numbers. I first discovered this in the multiplication table. What I did was to take the multiple of a number and then reduce that number down to a single digit. Reducing a number means that you add the digits of a number together until you end up with a single-digit number. Usually you only have to do this once, but occasionally you will have to reduce a number twice. If we were to reduce the number 16, we would do this: $1 + 6 = 7$. A more complex situation could occur as in reducing the number 48: $4 + 8 = 12$ and then $1 + 2 = 3$. In the second instance, we have to reduce the number twice.

Let's take a look at the first eight multiples of the number 8. We have 8, 16, 24, 32, 40, 48, 56, and 64. When we reduce those numbers we end up with 8, 7, 6, 5, 4, 3, 2, and 1. So what if all the multiples of all the numbers have these hidden patterns? That would be pretty amazing! Well, I'm here to tell you that they do all have patterns, but that's just the tip of the iceberg. Be prepared to have your mind blown, because I'm about to reveal all kinds of hidden patterns and sequences in numbers.

Reducing numbers is commonly found in numerology. Probably a lot of people think that numerology is just some crazy new-age woo-woo. Actually, it is a science that was discovered by Pythagoras over 2,500 years ago. He felt that numbers were the basis of the universe. From my explorations with numbers, I feel that numbers reveal a definite order to the universe. You can make your own conclusions on the subject, but all the patterns that emerge are way beyond the realm of coincidence.

I decided to start writing down some of the things I had discovered with numbers, and the more I played with the numbers, the more patterns I found. I felt like Alice going down the rabbit hole. A lot of mysteries were revealed to me, and I'm sure there are still a lot more to be found. My ultimate purpose in writing this book is to help kids who are struggling with math. This book reveals a whole new hidden dimension that will help to engage children who are having trouble with math, as well as those children who just don't like math. Most of the math in this book is suitable for about a fourth-grade-level student. The math involved is basically addition, subtraction, and multiplication. There is a section on exponents, which is generally taught in the higher grades, but some of it could be incorporated into elementary math. This information is also useful for younger students in second- or third-grade level.

The purpose of this book is to delight, engage, and excite young learners, as well as to inspire those who teach them; this book is not designed to regurgitate the Common Core. There are instances where the mysteries behind the patterns can and will be solved. Sometimes, I will discuss that and reveal the mysteries, and other times I will not in order to retain the air of mystery and excitement. If the students can uncover the mysteries themselves, they will be more engaged and excited, which is the whole goal of this book. This material is designed to be used as a supplement to your current curriculum or as a fun side activity.

In several sections, I will give you ideas for lesson plans to help your students or your children with their math skills. You can also go to my website (http://www.mindblowingmath.com) to find work sheets to use with your students. Options for different grades levels will also be discussed. There are also 20 pages of worksheets at the end of this book.

PREFACE

I'm sure you've seen these table dozens or maybe even hundreds of times, but did you know about all the patterns and mysteries to be found within? Let me show you what I've found. Be warned: it might blow your mind.

×	1	2	3	4	5	6	7	8	9	10
1	1	2	3	4	5	6	7	8	9	10
2	2	4	6	8	10	12	14	16	18	20
3	3	6	9	12	15	18	21	24	27	30
4	4	8	12	16	20	24	28	32	36	40
5	5	10	15	20	25	30	35	40	45	50
6	6	12	18	24	30	36	42	48	54	60
7	7	14	21	28	35	42	49	56	63	70
8	8	16	24	32	40	48	56	64	72	80
9	9	18	27	36	45	54	63	72	81	90
10	10	20	30	40	50	60	70	80	90	100

If we drop the multiples of 10, the multiplication table will look like this. I've color-coded the odd and even numbers, which gives us our first pattern. All the multiples of even number will give even results, while the multiples of the odd numbers will alternate between even and odd numbers. Fifty-six of the numbers are even, and twenty-five are odd.

×	1	2	3	4	5	6	7	8	9
1	1	2	3	4	5	6	7	8	9
2	2	4	6	8	10	12	14	16	18
3	3	6	9	12	15	18	21	24	27
4	4	8	12	16	20	24	28	32	36
5	5	10	15	20	25	30	35	40	45
6	6	12	18	24	30	36	42	48	54
7	7	14	21	28	35	42	49	56	63
8	8	16	24	32	40	48	56	64	72
9	9	18	27	36	45	54	63	72	81

If we draw a diagonal line from the upper left to the lower right, we would have symmetry, which is not surprising. We could fold the table along this line, and the same numbers would be on top of one another.

×	1	2	3	4	5	6	7	8	9
1	1	2	3	4	5	6	7	8	9
2	2	4	6	8	10	12	14	16	18
3	3	6	9	12	15	18	21	24	27
4	4	8	12	16	20	24	28	32	36
5	5	10	15	20	25	30	35	40	45
6	6	12	18	24	30	36	42	48	54
7	7	14	21	28	35	42	49	56	63
8	8	16	24	32	40	48	56	64	72
9	9	18	27	36	45	54	63	72	81

Could we possibly have symmetry if we draw the other diagonal? I am going to prove that we will. That's going to be in a later chapter, but before we get there, let's look at some other interesting things.

×	1	2	3	4	5	6	7	8	9
1	1	2	3	4	5	6	7	8	9
2	2	4	6	8	10	12	14	16	18
3	3	6	9	12	15	18	21	24	27
4	4	8	12	16	20	24	28	32	36
5	5	10	15	20	25	30	35	40	45
6	6	12	18	24	30	36	42	48	54
7	7	14	21	28	35	42	49	56	63
8	8	16	24	32	40	48	56	64	72
9	9	18	27	36	45	54	63	72	81

Start in the bottom left-hand corner and look at the numbers diagonally. First we have 9, and then 8 and 18, and then 7, 16 and 27, and next is 6, 14, 24 and 36. Can you see how the underlined numbers in the ones' place are symmetrical? It continues all the way to the opposite corner. So this is just the beginning. If you like math, you're in for a real treat. If you don't like math, you may change your mind after reading this book.

CHAPTER 1

The Numbers 0 through 9

Our numbering system is called a base-ten system. Most of the cultures around the world use this system. Our place values are: ones, tens, hundreds, thousands, and so on. We call each individual number a digit. Is it a coincidence that we also call our fingers and toes digits and that we have ten of each? Many other languages share this same coincidence. History is very unclear as to how or when numbers came to be. There is evidence of counting some twenty thousand years ago, but there is no record of numbers until about 4,000 BC in Samaria.

One thing is certain: numbers will never go away; they are here to stay. Besides math, we need numbers for our phones, our street addresses, our credit card numbers, and our social security numbers. We need numbers to figure out which clothes and shoes will fit us, not to mention that we need numbers for money to put values on goods and services.

Because this work revolves around reducing numbers, we are really only working with the numbers 1 through 9. The number 10 would reduce back to 1, as $1 + 0 = 1$. Any number, no matter how large, will reduce back to a single digit. Larger numbers will require more work, however, as we may need to reduce the number several times.

The Number 1

The number 1 is a single, solitary thing. It's a beginning, a starting point. In geometry, it would be a point. Counting by one is the first skill we learn in math. First we learn to count to 10, and then we learn to count to larger and larger numbers, eventually counting all the way up to 100.

My first discovery was from looking at multiples of numbers and then reducing those numbers. As was previously explained, reducing numbers means that you add the digits of a number together to get a single-digit number. For example, if your number is 36, you would add 3 + 6, which would equal 9 (3 + 6 = 9). There are occasions where you might have to reduce a number more than once. For example: 38 would give you 3 + 8 = 11, and then you would add 1 +1 = 2, so 38 would reduce to 2.

All numbers reduce down to a single digit or to the numbers 1 through 9. After we count up to 9, the next number, 10 (1 + 0), would reduce to 1. The number 11 (1 + 1) would reduce to 2, and so on.

After the number 9, the numbers start over at number 1 and continue on sequentially counting up to 9 and then repeating the process. Let me show you what that looks like in the chart below.

1	2	3	4	5	6	7	8	9
10	11	12	13	14	15	16	17	18
19	20	21	22	23	24	25	26	27
28	29	30	31	32	33	34	35	36
37	38	39	40	41	42	43	44	45
46	47	48	49	50	51	52	53	54
55	56	57	58	59	60	61	62	63
64	65	66	67	68	69	70	71	72

So how are all the numbers in the columns similar? They all reduce down to the number, which is at the top of the column. If we look at the numbers in column 5 as an example, we have 1 + 4, 2 + 3, 3 + 2, 4 + 1, 5 + 0, 5 + 9, and 6 + 8. The last two are a little more complex and require two reductions: 5 + 9 = 14 and then 1 + 4 = 5, and 6 + 8 = 14 and then 1 + 4 = 5.

Now let's take one of these numbers and look at their multiples and their reductions:

Number	Reduction/s	Final Answer
35	3 + 5	8
70	7 + 0	7
105	1 + 0 + 5	6
140	1 + 4 + 0	5
175	1 + 7 + 5 / 1 + 3	4
210	2 + 1 + 0	3
245	2 + 4 + 5 / 1 + 1	2
280	2 + 8 + 0 / 1 + 0	1
315	3 + 1 + 5	9

Since 35 can be reduced to 8, the pattern we get for the multiples of 35 is the same as the pattern we got for the multiples of 8. The number 35 is a multiple of 7 and 5, and all the multiples end in a 5 or a 0, but again, since 3 + 5 = 8, 35 shares the same pattern that we found with the multiples of 8.

Before we move on, let's add up all the numbers 1 through 9, which I will call the counting numbers for the number 1: 1 + 2 + 3 + 4 + 5 + 6 + 7 + 8 + 9 = 45. When we reduce 45, or add 4 + 5, we get 9.

Let me show you a nice trick for adding the numbers 1 through 9 together.

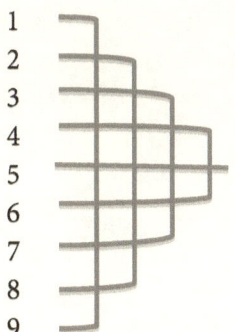

I find it helpful to connect the numbers that add up to 10: 1 + 9, 2 + 8, 3 + 7, and 4 + 6. There are four pairs of numbers that add up to 10. Then we have the number 5 left in the middle. Multiplying 4 × 10 = 40, and adding 40 + 5 = 45. Then we can reduce 45 by adding 4 + 5, which equals 9.

Teaching children to group numbers together for addition is a very valuable skill. Also showing children that numbers can be added in any order is a valuable and helpful tool. Children get very frustrated or overwhelmed when they're asked to add several numbers together. Using tools to make the task easier will greatly benefit your kids or students.

A worksheet on this lesson is included in the back of this book.

Class project: Ask your students to pick a number from 2 to 9. Next, ask them to write down that number and the next eight consecutive numbers. Now, ask the students to add all those numbers together. If they did it right, they will have a multiple of 9. Finally ask them to reduce their number to a single digit. For an added bonus, see if anyone can tell you why any nine consecutive numbers will reduce back to the number 9.

The Number 2

Two items make a pair. Lots of things come in pairs. Two is also used as a counting number. Counting by twos can make counting larger numbers easier. In geometry, the number 2 represents two points, which can be connected to form a line.

So we're going to explore the multiples of 2 and the reduction of the multiples of the number 2. Once again, to reduce a number, you add its digits together so that you get a single number. If we were to reduce 12, we would add 1 + 2 and that would equal 3 (1 + 2 = 3). As previously stated, sometimes you need to reduce a number twice, as in 58, where 5 + 8 = 13. You would then need to add 1 + 3 to get 4.

So let's look at the multiples of 2 and their reductions.

Number	Reduction	Final Answer
2	2	2
4	4	4
6	6	6
8	8	8
10	1 + 0	1
12	1 + 2	3
14	1 + 4	5
16	1 + 6	7
18	1 + 8	9

If we continue, the pattern repeats itself:

20	2 + 0	2
22	2 + 2	4
24	2 + 4	6

So we can see the pattern of counting up by even numbers and then counting up by odd numbers. All the numbers 1 through 9 are represented when we reduce the multiples of 2.

What if we skip up to a big number?

2000	2	2
2002	2 + 2	4
2004	2 + 4	6
2006	2 + 6	8
2008	2 + 8 = 10 (1 + 0 = 1)	1
2010	2 + 1 = 3	3
2012	2 + 1 + 2 = 5	5
2014	2 + 1 + 4 = 7	7
2016	2 + 1 + 6 = 9	9

Once again, we have the same sequence of numbers. This pattern will continue infinitely.

I have to note also that the odd numbers 1 + 3 + 5 + 7 + 9 would add up to 25, and then 2 + 5 reduces to 7.

The even numbers 2 + 4 + 6 + 8 add up to 20, and 2 + 0 = 2. If we add the 7 + 2, we get 9, which is the most magical number of all. If we didn't reduce the 25 and 20 first but instead added them together, we would have 20 + 25 = 45, which again reduces to 9.

Finally, let's add all the multiples of 2 together. Once again, there is an easy way to do this:

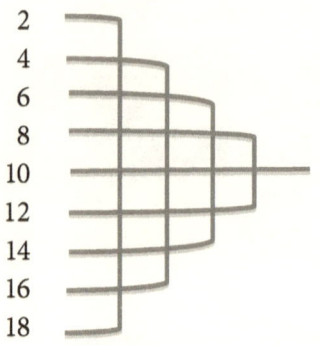

This time, we can connect numbers that add up to 20: 2 + 18, 4 + 16, 6 + 14, and 8 + 12. Then we have 10 left over in the middle. Multiplying 4 × 20 = 80. Adding 10 + 80 = 90, which reduces to 9.

The Number 3

The number 3 is an interesting number. In geometry, three points form a triangle, our first two-dimensional figure. A triangle makes for a very strong and sturdy shape. It is used a lot in building bridges and trusses. The multiples of 3 when reduced form a pattern using just three numbers. Can you guess what they might be? The pattern formed is very similar to another number. What number do you think is related to 3? Let's look at the multiples of 3:

Number	Reduction	Final Answer
3	3	3
6	6	6
9	9	9
12	1 + 2	3
15	1 + 5	6
18	1 + 8	9
21	2 + 1	3
24	2 + 4	6
27	2 + 7	9

It keeps repeating infinitely. What do we get if we add and reduce 3, 6, and 9? 3 + 6 + 9 = 18, so we have to reduce again: 1 + 8 = 9. Stay tuned for the amazing number 9—it's pretty special!

Let's add all the multiples of 3 together. Again, here is an easy way to do that:

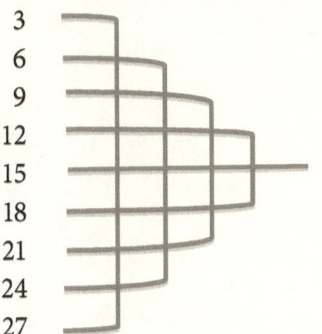

3
6
9
12
15
18
21
24
27

An easy way to do this is to add: 3 + 27, 6 + 24, 9 + 21, and 12 + 18. That will give us four 30s and a 15 left in the middle. Multiplying $4 \times 30 = 120$. When we add the 15 to this sum, we get 135. Let's reduce: $1 + 3 + 5 = 9$. We get 9 again—imagine that!

What happens if we multiply any combination of 3, 6, and 9 together? Give it a try. It's a little spooky.

$3 \times 3 = 9$
$3 \times 6 = 18 \ (1 + 8 = 9)$
$3 \times 9 = 27 \ (2 + 7 = 9)$
$6 \times 6 = 36 \ (3 + 6 = 9)$
$6 \times 9 = 54 \ (5 + 4 = 9)$
$9 \times 9 = 81 \ (8 + 1 = 9)$

The numbers 3, 6, and 9 definitely like to hang out together. You will notice that in other sections, as well. It makes sense that they would do that, as 3 divides the number 9 evenly. Then 6 is like two 3s, and 9 is three 3s.

There's another number that is very similar to the number 3. Can you guess what it is?

The Number 4

Okay, so you're probably saying, what's so special about the number 4? Let's take a look and see. A 4 is two pairs, or two 2s. In geometry, 4 can be a square, rectangle, rhombus, kite, trapezoid, parallelogram, or some other quadrilateral (four-sided figure). The number 4 is, in fact, a perfect square of 2×2. It's the only number that gets the same results by adding it to itself or multiplying it by itself: $2 \times 2 = 4$, and $2 + 2 = 4$. Let's look at the multiples and reductions of the number 4.

Number	Reduction	Final Answer
4	4	4
8	8	8
12	1 + 2	3
16	1 + 6	7
20	2 + 0	2
24	2 + 4	6
28	2 + 8 = 10 (1 + 0 = 1)	1
32	3 + 2	5
36	3 + 6	9

Do you see the pattern? This pattern is a little more complex. There are actually two patterns that alternate here: 4, 3, 2, 1, 9 and 8, 7, 6, 5. All the numbers 1 through 9 are used in these patterns. When we add the numbers 4, 3, 2, 1, and 9 together, we get 19, which reduces to 1 ($1 + 9 = 10$, and then $1 + 0 = 1$). If we add all the numbers from the second pattern—8, 7, 6, and 5—we get 26, which reduces to 8 ($2 + 6 = 8$). Finally, $1 + 8 = 9$.

Looking at the numbers in the ones' place, we have a repeating pattern of 4, 8, 2, 6, 0, 4, 8, 2, 6. This will keep repeating indefinitely.

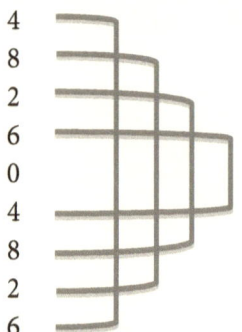

4
8
2
6
0
4
8
2
6

If we add these numbers together—$4 + 8 + 2 + 6 + 4 + 8 + 2 + 6$—we come up with 40, which reduces to 4 ($4 + 0$). Numbers are magic! They actually combine nicely into four 10s—$4 + 6$ and $2 + 8$ both occur twice, which gives us the four 10s.

Adding all the numbers in the tens' place together—$1 + 1 + 2 + 2 + 2 + 3 + 3 = 14$, reduces to 5 ($1 + 4 = 5$), and then $4 + 5 = 9$. The number 9, as you will learn, is very magical.

If we add all the multiples of 4 together, we would come up with 180, which reduces to 9 (1 + 8 = 9).

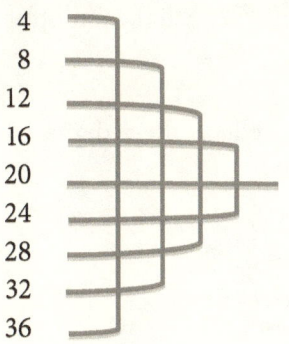

We can add them easily if we make four pairs of 40: 4 + 36, 8 + 32, 12 + 28, and 16 + 24. We are left with a 20 in the middle, which is half of 40. Multiplying 4 × 40 = 160, and then when we add the 20, we get 180. Then 1 + 8 + 0 reduces to that magical number, 9.

The Number 5

Is there any magic to be found in the number 5? The number 5 is right in the middle of the numbers 1 through 9. After learning to count by ones and twos, counting by fives is very easy. The number 5 is important in rounding, telling us whether to go up or stay the same. The number 5 is half of 10, which is the base for our number system. A pentagon and five-pointed star are some geometric shapes associated with the number five. The number 5 is also a prime number, because it is only divisible by 1 and itself. Is there anything special about the multiples of 5 and their reductions? Let's take a look:

Number	Reduction	Final Answer
5	5	5
10	1 + 0	1
15	1 + 5	6
20	2 + 0	2
25	2 + 5	7
30	3 + 0	3
35	3 + 5	8
40	4 + 0	4
45	4 + 5	9

Do you see a pattern? This is similar to the pattern with the number 4. There are actually two alternating patterns once again. This time, we are counting up: 5, 6, 7, 8, 9 and 1, 2, 3, 4. Again, all the numbers 1 through 9 are in the pattern. All the numbers add up to 45, which reduces to 9 (4 + 5 = 9). All the multiples of 5 end in either a 0 or a 5. The multiples of 5 alternate between even and odd numbers, which is true with the multiples of any odd number.

The 5s also have a special pattern when you reduce their digits by subtraction. Let me show you what I mean by that:

Number	Subtraction	Final Answer	
5	5 - 0	5	
10	1 - 0	1	
15	5 - 1	4	
20	2 - 0	2	
25	5 - 2	3	
30	3 - 0	3	
35	5 - 3	2	
40	4 - 0	4	
45	5 - 4	1	
50	5 - 0	5	

In order to achieve symmetry here with this pattern, I had to continue up to 5 × 10, or 50.

Finally, let's add up all the multiples of 5. Once again, there's an easy way to do that:

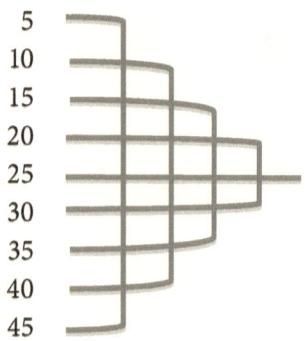

Adding all the multiples of 5 gives us 225, which would reduce to 9 (2 + 2 + 5 = 9). We get four pairs that add up to 50: 5 + 45, 10 + 40, 15 + 35, and 20 + 30. That leaves us with 25 left in the middle. Multiplying 4 × 50 = 200, which, added to 25, gives us 225. Like all the previous reductions of multiples, the numbers 2 + 2 + 5 equal our favorite number, 9!

The Number 6

Now we're back to another even number. The number 6 is very similar to another number. Which number do you think that might be? The number 6 is a multiple of 2 and 3. In geometry, a six-sided figure is called a hexagon. I find it interesting that *hex* is the Greek word for *six*. The number 6 has bad connotations and good connotations, depending on what culture you look at. The Bible considers it to be a number that recognizes the weakness of man. On the other hand, the Chinese consider the number 6 to be a lucky number. In numerology, the number 6 represents nurturing. They consider it to be a number that gets along with all other numbers. Let's see what patterns can be found with the magical number 6.

Number	Reduction	Final Answer
6	6	6
12	1 + 2	3
18	1 + 8	9
24	2 + 4	6
30	3 + 0	3
36	3 + 6	9
42	4 + 2	6
48	4 + 8 = 12 (1 + 2)	3
54	5 + 4	9

So we have a repeating 6, 3, and 9 pattern. If you guessed that the pattern was similar to the pattern found with the number 3, you would be correct. That pattern was 3, 6, and 9 repeating. All the multiples of the number 6 are even, but most of the reductions are odd. If we add the 6, 3, and 9 together, we get 18, which reduces to 9 (1 + 8 = 9). Let's add all the reductions together: three 6s are 18 (which reduces to 9), three 3s equal 9, three 9s equal 27, which also reduces to 9. Then 18 + 9 + 27 = 54, which also reduces to 9. As we saw with the number 3, multiplying all of the combinations of 3, 6, and 9 together will give us the number 9 when reduced.

$3 \times 3 = 9$
$3 \times 6 = 18 \ (1 + 8 = 9)$
$3 \times 9 = 27 \ (2 + 7 = 9)$
$6 \times 6 = 36 \ (3 + 6 = 9)$

$6 \times 9 = 54 \ (5 + 4 = 9)$
$9 \times 9 = 81 \ (8 + 1 = 9)$

Let's add all the multiples of 6 together and see what we come up with. Here's an easy way to do that:

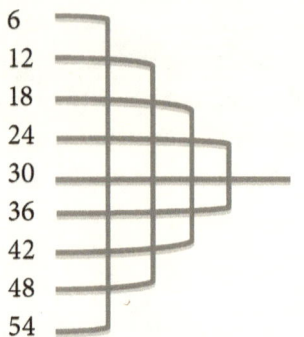

6
12
18
24
30
36
42
48
54

We have four pairs of numbers that add up to 60, and we are left with a 30 in the middle. Multiply $4 \times 60 = 240$, which added to 30 equals 270. Then, magically, $2 + 7 + 0 = 9$.

The Number 7

The number 7 has always been a favorite of mine. It seems special, magical, maybe even lucky. If you put it together with 11, it's very lucky. There are seven days of the week—that's pretty special. The musical scale is composed of seven notes. When rolling two dice, you have the highest probability of rolling a seven. The number 7 is odd and prime. Is there any magic to be found in the multiples of 7? Let's take a look.

Number	Reduction	Final Answer
7	7	7
14	1 + 4	5
21	2 + 1	3
28	2 + 8 = 10 (1 + 0)	1
35	3 + 5	8
42	4 + 2	6
49	4 + 9 = 13 (1 + 3)	4
56	5 + 6 = 11 (1 + 1)	2
63	6 + 3	9

There is a pattern of counting down by odd numbers and then counting down by even numbers. Once again, all the numbers 1 through 9 are used in this pattern. As we've seen before, the numbers 1 through 9 add up to 45, which reduces to 9. Let's add up all the number in the ones' place: 7 + 4 + 1 + 8 + 5 + 2 + 9 + 6 + 3 = 45. Add 4 + 5 = 9. Now we'll add up all the numbers in the tens' place: 1 + 2 + 2 + 3 + 4 + 4 + 5 + 6 = 27. Add 2 + 7 = 9. We could continue and add the 9 + 9 together to get 18, which also reduces to 9.

There's another pattern I see with the numbers in the ones' place for the multiples of 7.

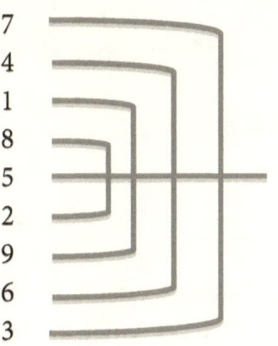

I happened to notice that we have four pairs of numbers that add up to 10: 7 + 3, 4 + 6, 1 + 9, and 8 + 2. Then we are left with a 5 in the middle. Multiply 4 × 10 = 40, and adding 5 equals 45, which magically reduces to 9.

What happens when we add all the multiples of 7 together?

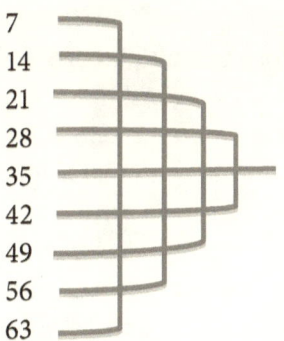

Once again, there is an easy way to add all these numbers together. This time, we have four sets of numbers that add up to 70: 7 + 63, 14 + 56, 21 + 49, and 28 + 42. Then we are left with 35 in the middle. Multiply 4 × 70 = 280. When we add 35 + 280, we get 315, and 3 + 1 + 5 = 9.

The Number 8

The number 8 is yet another magical number. The shape of 8 is interesting, with its two loops. If we turn an 8 on its side, we have the infinity symbol. An octagon and an octopus come to mind when thinking about the number 8. I'm not sure what happened with October, as it is the tenth month. A quick Internet search tells me that October was originally the eighth month in the Roman calendar. Later on, January and February were added to the calendar, but October kept its original name. The number 8 is an even number, and it is also our first perfect cube: $2 \times 2 \times 2 = 8$. Let's see what mysteries we can uncover with the number 8.

Number	Reduction	Final Answer
8	8	8
16	$1 + 6$	7
24	$2 + 4$	6
32	$3 + 2$	5
40	$4 + 0$	4
48	$4 + 8 = 12 \ (1 + 2)$	3
56	$5 + 6 = 11 \ (1 + 1)$	2
64	$6 + 4 = 10 \ (1 + 0)$	1
72	$7 + 2$	9

So again, we find a pattern in our reductions. This time, we are counting down from 8. Again, all the numbers 1 through 9 are represented in the pattern. We already know that if we add the numbers 1 through 9, we come up with 45, which reduces to 9. All the multiples of 8 are even numbers, which is true for all multiples of even numbers.

Let's try adding up our ones' and tens' columns for the multiples of 8. In the tens' places, we have: $1 + 2 + 3 + 4 + 4 + 5 + 6 + 7 = 32$. Adding $3 + 2$ reduces to 5 (3+2=5). In the ones' column, we have $8 + 6 + 4 + 2 + 8 + 6 + 4 + 2 = 40$, and $4 + 0 = 4$. Finally, adding $5 + 4$, we come up with that magic number 9 ($5 + 4 = 9$).

Here's another little interesting pattern with 8s. Let's look at the tens' and ones' digits for the multiples of 8: 8, 16, 24, 32, 40, 48, 56, 64, and 72.

Tens

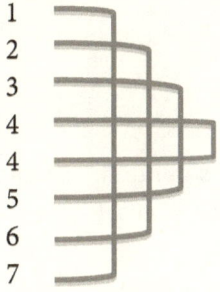

1
2
3
4
4
5
6
7

I find it interesting how the numbers connect to add up to 8. We have four sets of numbers that add up to 8: 1 + 7, 2 + 6, 3 + 5, and 4 + 4.

Ones

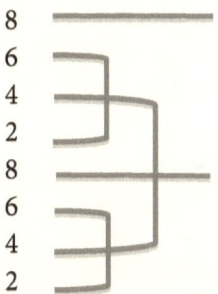

8
6
4
2
8
6
4
2

With the ones, we have three sets of numbers that add up to 8: 6 + 2, 4 + 4, and 6 + 2. Then we have two numbers that are already 8, for a total of five 8s. If we add those five to the four 8s we got from adding the tens' digits, we'd have nine 8s.

There is also a pattern if we subtract the digits of the multiples of 8.

Multiple	Subtraction	Solution
8	8 - 0	8
16	6 - 1	5
24	4 - 2	2
32	3 - 2	1
40	4 - 0	4
48	8 - 4	4
56	6 - 5	1
64	6 - 4	2
72	7 - 2	5
. . . .		
80	8 - 0	8

So we have a pattern: 8, 5, 2, 1, 4, 4, 1, 2, 5, 8—although in this case for the pattern to be symmetrical, we have to continue up to 8 × 10, which equals 80.

There is another interesting pattern I see with the last set of numbers we came up with from subtracting the digits of the multiples of 8.

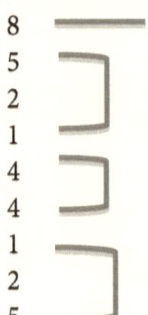

8
5
2
1
4
4
1
2
5

Now we have four groups of numbers that add up to 8: 8, 5 + 2 + 1, 4 + 4, and 1 + 2 + 5.

Let's take a look at what all the multiples of 8 add up to.

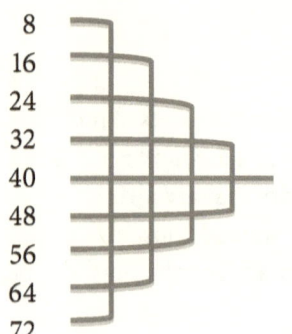

8
16
24
32
40
48
56
64
72

There are four pairs of numbers that add up to 80: 8 + 72, 16 + 64, 24 + 56, and 32 + 48. That leaves us with 40 left in the middle. Multiply 4 × 80 = 320, and add 320 + 40 = 360. It should be of no surprise to you that 360 reduces to 9 (3 + 6 + 0 = 9).

There is one more interesting pattern I have found with the multiples of 8. If we add the digits in the ones' and tens' places together in the smaller number, the result will be the tens' digit of the larger number. Let me explain it with the following chart:

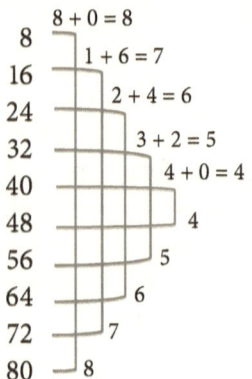

8   8 + 0 = 8
16   1 + 6 = 7
24   2 + 4 = 6
32   3 + 2 = 5
40   4 + 0 = 4
48   4
56   5
64   6
72   7
80   8

The Number 9

The number 9 is the most magical of all the numbers. I think this may be because it contains all the other numbers. The number 9 is the sum when adding 1 + 8, 2 + 7, 3 + 6, or 4 + 5. When any other number is added to 9, it reduces to itself. Here are some examples 4 + 9 = 13, (1 + 3 = 4) and 7 + 9 = 16 (1 + 6 = 7). When working with a larger number, if you reduce that number first, as in 33 (which reduces to 3 + 3 = 6), and then 6 + 9 = 15 (1 + 5 = 6), you will get the same result by reducing after adding the numbers 33 + 9 = 42 (4 + 2 = 6). Also, as we have seen before, if you add the numbers 1 through 9, the result is 45, which also reduces to the amazing number 9.

Lots of magic occurs when you multiply by 9, as well. Let's looks at the multiples of 9:

9 × 1 = 09
9 × 2 = 18
9 × 3 = 27
9 × 4 = 36
9 × 5 = 45
9 × 6 = 54
9 × 7 = 63
9 × 8 = 72
9 × 9 = 81
9 × 9 = 90

If we look at the ones' digits, we can see the numbers decreasing from 9 to 0, while the numbers in the tens' digits increase from 0 to 9. Then we have a lot of pairs of reverse digit numbers: 09–90, 18–81, 27–72, 36–63, and 45–54. Finally, all these numbers reduce to 9: 1 + 8, 2 + 7, 3 + 6, and 4 + 5 all equal 9. What happens with larger numbers? Let's multiply 9 × 342. That would equal 3,078. When we reduce that number, 3 + 0 + 7 + 8 = 18, so we have to reduce again: 1 + 8 = 9. So then the opposite is also true. Any number that reduces to 9 is divisible by 9. Let's try that with 342: 342 ÷ 9 = 38. We can't forget that 9 is also a perfect square, so anything that is divisible by 9 is also divisible by 3. I also find it interesting that there are three sets of three numbers that add up to 9:

126
135
234

There are three groups of three, for a total of nine numbers. Numbers are magic!

There is also a pattern when we subtract the digits of the multiples of 9.

9 - 0 = 9
8 - 1 = 7
7 - 2 = 5
6 - 3 = 3
5 - 4 = 1
5 - 4 = 1
6 - 3 = 3
7 - 2 = 5
8 - 1 = 7
9 - 0 = 9

It's interesting to note that all these numbers are odd. Maybe I'm stretching it a bit, but these numbers we get by subtracting (9, 7, 5, 3, 1, 1, 3, 5, 7, 9) add up to 50, which reduces to 5. This is also the mean, or average, for this group of numbers.

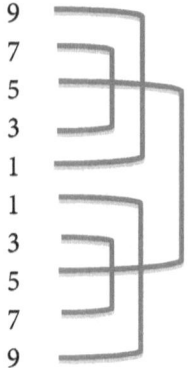

9
7
5
3
1
1
3
5
7
9

Here's an interesting way to add up the five sets of 10: 9 + 1 and 7 + 3 both occur twice. Then we also have 5 + 5.

What happens when we add up the multiples of 9?

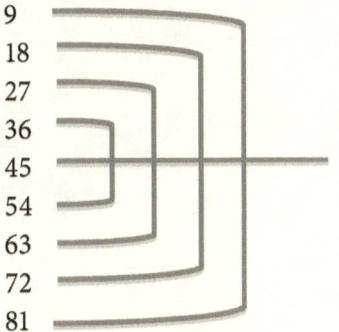

9
18
27
36
45
54
63
72
81

Adding up the multiples of 9 gives us four pairs of numbers that add up to 90: 9 + 81, 18 + 72, 27 + 63, and 36 + 54. We are left with 45 in the middle. Multiply 4 × 90 = 360, which added to 45 equals 405, which magically reduces to the incredible number 9!

The Lowly Number 0

The number that has not been mentioned so far is the lowly number 0. The 0 doesn't really play a part in this number magic. Adding or subtracting 0 does not change the value of a number. If we add 4 + 0, we end up with 4, and 4 - 0 also equals 4. Multiplying by 0 turns everything into 0. Some examples would be these: 5 × 0 = 0 and 2,987,654 × 0 = 0. You can forget about dividing by 0, as that just can't be done. If you think about it, you can't divide something by nothing. A 0 is nothing, although there are occasions where 0 does have some value. We'll explore that later, where 0 is important is as a placeholder. How could we have the number 1001 without 0? The numbers 1,000,000 or 0.0001 (one ten-thousandth) would not be possible without the 0. It's interesting that 0 is shaped like a circle or an oval. It's basically an empty hole, but the shape works very well in making other numbers work. The number 0 deserves a lot of credit for being a good placeholder.

CHAPTER 2

Recap and Summary of the Numbers 1 through 9

This table shows what all the multiples of the individual numbers add up to, along with their reductions.

×	1	2	3	4	5	6	7	8	9	Totals	Reductions
1	1	2	3	4	5	6	7	8	9	45	$4 + 5 = 9$
2	2	4	6	8	10	12	14	16	18	90	$9 + 0 = 9$
3	3	6	9	12	15	18	21	24	27	135	$1 + 3 + 5 = 9$
4	4	8	12	16	20	24	28	32	36	180	$1 + 8 + 0 = 9$
5	5	10	15	20	25	30	35	40	45	225	$2 + 2 + 5 = 9$
6	6	12	18	24	30	36	42	48	54	270	$2 + 7 + 0 = 9$
7	7	14	21	28	35	42	49	56	63	315	$3 + 1 + 5 = 9$
8	8	16	24	32	40	48	56	64	72	360	$3 + 6 + 0 = 9$
9	9	18	27	36	45	54	63	72	81	405	$4 + 0 + 5 = 9$

As you can see, they all add up to a multiple of 9 and finally reduce to 9. Going from one total to the next, we increase by 45 each time, and since 45 reduces to a 9, each total must therefore also reduce to 9. So this mystery can be solved easily, but don't tell your students. Either leave it as a mystery or let them figure it out for themselves. Both of those scenarios will make the learning process more fun for them.

This table shows what all the reduced multiples add up to and then reduce to. As you can see, they all add up to a multiple of 9 and finally reduce to 9, just like the numbers in the previous table reduced to 9.

×	1	2	3	4	5	6	7	8	9	Totals	Reductions
1	1	2	3	4	5	6	7	8	9	45	9
2	2	4	6	8	1	3	5	7	9	45	9
3	3	6	9	3	6	9	3	6	9	54	9
4	4	8	3	7	2	6	1	5	9	45	9
5	5	1	6	2	7	3	8	4	9	45	9
6	6	3	9	6	3	9	6	3	9	54	9
7	7	5	3	1	8	6	4	2	9	45	9
8	8	7	6	5	4	3	2	1	9	45	9
9	9	9	9	9	9	9	9	9	9	81	9

I find it interesting to note that the reductions for the numbers 1, 2, 4, 5, 7, and 8 all add up to 45, while the reductions for the numbers 3 and 6 add up to 54. The number 9 is much greater than all the others by adding up to 81. All the numbers become equal again, as they all reduce to 9.

I wanted to see if I could find any patterns by highlighting the even and odd numbers of the reductions, and I do see some interesting surprises. But wait until you see the next table.

×	1	2	3	4	5	6	7	8	9
1	1	2	3	4	5	6	7	8	9
2	2	4	6	8	1	3	5	7	9
3	3	6	9	3	6	9	3	6	9
4	4	8	3	7	2	6	1	5	9
5	5	1	6	2	7	3	8	4	9
6	6	3	9	6	3	9	6	3	9
7	7	5	3	1	8	6	4	2	9
8	8	7	6	5	4	3	2	1	9
9	9	9	9	9	9	9	9	9	9

Now I took out the row and column of 9s, and it gets really interesting. I also removed the top row and the left-hand column. We have diagonal symmetry in both directions! I did promise this to you in the beginning of the book, and I must say that I find this truly amazing. Notice also how row 1 and row 8 are exact opposites. The same is true for rows 2 and 7, 3 and 6, and even for rows 4 and 5, or all the pairs of number that add up to 9. Also, every pair of corresponding numbers in the paired row (2–7, 3–6, and 4–5) all add up to 9. We also have symmetry among all the rows and columns of numbers if we flip one of the rows or columns. There are two instances where we add 9 + 9, which equals 18, but 18 of course reduces to 9. Thirty of the numbers are even, and thirty-four are odd. I've decided to name this Nansea's Square. I figure if Pascal can put his name on a triangle, I can put my name on a square.

Nansea's Square

1	2	3	4	5	6	7	8
2	4	6	8	1	3	5	7
3	6	9	3	6	9	3	6
4	8	3	7	2	6	1	5
5	1	6	2	7	3	8	4
6	3	9	6	3	9	6	3
7	5	3	1	8	6	4	2
8	7	6	5	4	3	2	1

If we color-code the different numbers, patterns emerge. The number that shows up the least is 9, and the numbers that occur most often are 3 and 6. The numbers 1, 2, 4, 5, 7, and 8 each occur six times. There is a repeating pattern at least up through 8 of the frequency of the individual numbers, as you will see in the next chart.

1	2	3	4	5	6	7	8
2	4	6	8	1	3	5	7
3	6	9	3	6	9	3	6
4	8	3	7	2	6	1	5
5	1	6	2	7	3	8	4
6	3	9	6	3	9	6	3
7	5	3	1	8	6	4	2
8	7	6	5	4	3	2	1

Number	Frequency
1	6
2	6
3	12
4	6
5	6
6	12
7	6
8	6
9	4

Notice the repeating pattern under frequency of 6, 6, and 12. The number 6 seems to be the king here, as the frequency numbers are all 6 or a multiple of 6, and they all add up to 60, which reduces to 6—that is, if we ignore the 4 from the frequency of 9.

Here's some further proof that 6 is the king of the frequency chart we just looked at. If we take our number, multiply it by the frequency and then reduce that number, we will find a pattern we've seen before.

Number	Frequency	Number × Frequency	Reduction
1	6	6	6
2	6	12	3
3	12	36	9
4	6	24	6
5	6	30	3
6	12	72	9
7	6	42	6
8	6	48	3
9	4	36	9

Our reduction of the repeating 6, 3, and 9 is the same pattern we discovered from the multiples of 6. One other thing that should be noted is that all the numbers in the column "Number × Frequency" (6, 12, 36, 24, 30, 72, 42, 48, 36) are multiples of 6.

Since color-coding the whole table got a little muddy and confusing, I decided to show the individual numbers separately. The 9s make a nice little square pattern in the center of the grid, which happens to be in rows and columns that also include 3s and 6s. These three numbers like to stick together as has been noted before.

1	2	3	4	5	6	7	8
2	4	6	8	1	3	5	7
3	6	9	3	6	9	3	6
4	8	3	7	2	6	1	5
5	1	6	2	7	3	8	4
6	3	9	6	3	9	6	3
7	5	3	1	8	6	4	2
8	7	6	5	4	3	2	1

I grouped the 8s and 1s together, as they add up to 9, and they seem to share a pattern in an inverse or upside-down kind of way.

1	2	3	4	5	6	7	8
2	4	6	8	1	3	5	7
3	6	9	3	6	9	3	6
4	8	3	7	2	6	1	5
5	1	6	2	7	3	8	4
6	3	9	6	3	9	6	3
7	5	3	1	8	6	4	2
8	7	6	5	4	3	2	1

Color-coding the numbers individually also reveals some interesting patterns. Note how the smaller numbers form their shape from upper left to lower right, while the larger numbers go from lower left to upper right.

1	2	3	4	5	6	7	8
2	4	6	8	1	3	5	7
3	6	9	3	6	9	3	6
4	8	3	7	2	6	1	5
5	1	6	2	7	3	8	4
6	3	9	6	3	9	6	3
7	5	3	1	8	6	4	2
8	7	6	5	4	3	2	1

1	2	3	4	5	6	7	8
2	4	6	8	1	3	5	7
3	6	9	3	6	9	3	6
4	8	3	7	2	6	1	5
5	1	6	2	7	3	8	4
6	3	9	6	3	9	6	3
7	5	3	1	8	6	4	2
8	7	6	5	4	3	2	1

The 2s and 7s also are related and make a nice pattern, as well.

1	2	3	4	5	6	7	8
2	4	6	8	1	3	5	7
3	6	9	3	6	9	3	6
4	8	3	7	2	6	1	5
5	1	6	2	7	3	8	4
6	3	9	6	3	9	6	3
7	5	3	1	8	6	4	2
8	7	6	5	4	3	2	1

A similar pattern emerges with the 2s and 7s when they are highlighted individually.

1	2	3	4	5	6	7	8
2	4	6	8	1	3	5	7
3	6	9	3	6	9	3	6
4	8	3	7	2	6	1	5
5	1	6	2	7	3	8	4
6	3	9	6	3	9	6	3
7	5	3	1	8	6	4	2
8	7	6	5	4	3	2	1

1	2	3	4	5	6	7	8
2	4	6	8	1	3	5	7
3	6	9	3	6	9	3	6
4	8	3	7	2	6	1	5
5	1	6	2	7	3	8	4
6	3	9	6	3	9	6	3
7	5	3	1	8	6	4	2
8	7	6	5	4	3	2	1

The 3s and 6s which are the most plentiful number in the square make a nice little tic-tac-toe grid, leaving all the vertices open for the might number nine.

1	2	3	4	5	6	7	8
2	4	6	8	1	3	5	7
3	6	9	3	6	9	3	6
4	8	3	7	2	6	1	5
5	1	6	2	7	3	8	4
6	3	9	6	3	9	6	3
7	5	3	1	8	6	4	2
8	7	6	5	4	3	2	1

The 3s and 6s show a different pattern than what we saw with the 1s and 8s or 2s and 7s, but that seems to always be the way with 3s and 6s—they do their own thing in conjunction with 9.

1	2	3	4	5	6	7	8
2	4	6	8	1	3	5	7
3	6	9	3	6	9	3	6
4	8	3	7	2	6	1	5
5	1	6	2	7	3	8	4
6	3	9	6	3	9	6	3
7	5	3	1	8	6	4	2
8	7	6	5	4	3	2	1

1	2	3	4	5	6	7	8
2	4	6	8	1	3	5	7
3	6	9	3	6	9	3	6
4	8	3	7	2	6	1	5
5	1	6	2	7	3	8	4
6	3	9	6	3	9	6	3
7	5	3	1	8	6	4	2
8	7	6	5	4	3	2	1

The 4s and 5s also make a nice pattern, which is circular in shape.

1	2	3	4	5	6	7	8
2	4	6	8	1	3	5	7
3	6	9	3	6	9	3	6
4	8	3	7	2	6	1	5
5	1	6	2	7	3	8	4
6	3	9	6	3	9	6	3
7	5	3	1	8	6	4	2
8	7	6	5	4	3	2	1

When we separate the 4s and 5s, we see a similar pattern to most of the other numbers.

1	2	3	4	5	6	7	8
2	4	6	8	1	3	5	7
3	6	9	3	6	9	3	6
4	8	3	7	2	6	1	5
5	1	6	2	7	3	8	4
6	3	9	6	3	9	6	3
7	5	3	1	8	6	4	2
8	7	6	5	4	3	2	1

1	2	3	4	5	6	7	8
2	4	6	8	1	3	5	7
3	6	9	3	6	9	3	6
4	8	3	7	2	6	1	5
5	1	6	2	7	3	8	4
6	3	9	6	3	9	6	3
7	5	3	1	8	6	4	2
8	7	6	5	4	3	2	1

So then I had a thought. I wondered what would happen if we added diagonally across the multiplication table. Would we find a pattern there? For this exercise, I used the 8 × 8 multiplication board, which has the numbers already reduced. We could use the regular 8 × 8 multiplication board (without reducing those numbers), but that would just be more work. I'll show you what I mean by that a little later.

1	2	3	4	5	6	7	8
2	4	6	8	1	3	5	7
3	6	9	3	6	9	3	6
4	8	3	7	2	6	1	5
5	1	6	2	7	3	8	4
6	3	9	6	3	9	6	3
7	5	3	1	8	6	4	2
8	7	6	5	4	3	2	1

I color-coded the different diagonals to make them easier to see. Starting from the upper left, these are the numbers in the successive diagonals.

1
2 + 2
3 + 4 + 3
4 + 6 + 6 + 4
5 + 8 + 9 + 8 + 5
6 + 1 + 3 + 3 + 1 + 6
7 + 3 + 6 + 7 + 6 + 3 + 7
8 + 5 + 9 + 2 + 2 + 9 + 5 + 8
7 + 3 + 6 + 7 + 6 + 3 + 7
6 + 1 + 3 + 3 + 1 + 6
5 + 8 + 9 + 8 + 5
4 + 6 + 6 + 4
3 + 4 + 3
2 + 2
1

Hopefully you can see the symmetry in these rows of numbers, like 4, 6, 6, 4, for example. Notice also the patterns in the columns and also the back diagonals of these numbers.

FYI: I'm going to call this a positive reduction, since the numbers are going in a positive slope direction, which is basically from lower left to upper right. Okay—so let's add them up and see what we get:

Numbers	Totals	Reductions +
1	1	1
2 + 2	4	4
3 + 4 + 3	10	1
4 + 6 + 6 + 4	20	2
5 + 8 + 9 + 8 + 5	35	8
6 + 1 + 3 + 3 + 1 + 6	20	2
7 + 3 + 6 + 7 + 6 + 3 + 7	39	3
8 + 5 + 9 + 2 + 2 + 9 + 5 + 8	48	3
7 + 3 + 6 + 7 + 6 + 3 + 7	39	3
6 + 1 + 3 + 3 + 1 + 6	20	2
5 + 8 + 9 + 8 + 5	35	8
4 + 6 + 6 + 4	20	2
3 + 4 + 3	10	1
2 + 2	4	4
1	1	1

So indeed, we have patterns and repeating patterns and symmetry. There's even pattern and symmetry in most of the totals before we reduce the numbers (1, 4, 10 and 20, 35, 20 and 39, 48, 39). Don't you just love this number magic? Also, if we add and reduce the different patterns, we find something interesting: 1 + 4 + 1 = 6, 2 + 8 + 2 = 12, and 1 + 2 = 3, and then 3 + 3 + 3 = 9. So we have a 6, 3, 9 pattern again. Also, there are fifteen diagonals, and 1 + 5 = 6. I don't think it's all just coincidence. Take a look down the center column of the numbers being added. Do you see 1 + 4 + 9 + 7 + 7 + 9 + 4 + 1? You'll be seeing that pattern again.

If you want to work a little harder, you could add the numbers together straight from the multiplication table before they have been reduced. Let's give that a try:

×	1	2	3	4	5	6	7	8
1	1	2	3	4	5	6	7	8
2	2	4	6	8	10	12	14	16
3	3	6	9	12	15	18	21	24
4	4	8	12	16	20	24	28	32
5	5	10	15	20	25	30	35	40
6	6	12	18	24	30	36	42	48
7	7	14	21	28	35	42	49	56
8	8	16	24	32	40	48	56	64

Add 6 + 10 + 12 + 12 + 10 + 6 = 56, then add 5 + 6 = 11, and then add 1 + 1 = 2. This is the same answer we got with 6 + 1 + 3 + 3 + 1 + 6 = 20, and 2 + 0 = 2.

If you'd like a bigger challenge, you can add 8 + 14 + 18 + 20 + 20 + 18 + 14 + 8 = 120, and then add 1 + 2 + 0 = 3, which is the same as 8 + 5 + 9 + 2 + 2 + 9 + 5 + 8 = 48, 4 + 8 = 12, 1 + 2 = 3. Personally, I prefer adding the numbers after they've been reduced.

So is there anything special about the other diagonals? I'll call them the negative slope diagonals, or the ones that go from upper left to lower right. Let's check it out:

1	2	3	4	5	6	7	8
2	4	6	8	1	3	5	7
3	6	9	3	6	9	3	6
4	8	3	7	2	6	1	5
5	1	6	2	7	3	8	4
6	3	9	6	3	9	6	3
7	5	3	1	8	6	4	2
8	7	6	5	4	3	2	1

Again, I color-coded the different diagonals to make them easier to see. Starting from the upper right, they are:

8
7 + 7
6 + 5 + 6
5 + 3 + 3 + 5
4 + 1 + 9 + 1 + 4
3 + 8 + 6 + 6 + 8 + 3
2 + 6 + 3 + 2 + 3 + 6 + 2
1 + 4 + 9 + 7 + 7 + 9 + 4 + 1
2 + 6 + 3 + 2 + 3 + 6 + 2
3 + 8 + 6 + 6 + 8 + 3
4 + 1 + 9 + 1 + 4
5 + 3 + 3 + 5
6 + 5 + 6
7 + 7
8

There is symmetry in all the rows and columns and along the back diagonals. It's pretty crazy!

This reduction I'm going to call negative, as it represents a negative slope direction. Okay—so let's add them up and see what we get.

Numbers	Totals	Reductions –
8	8	8
7 + 7	14	5
6 + 5 + 6	17	8
5 + 3 + 3 + 5	16	7
4 + 1 + 9 + 1 + 4	19	1
3 + 8 + 6 + 6 + 8 + 3	34	7
2 + 6 + 3 + 2 + 3 + 6 + 2	24	6
1 + 4 + 9 + 7 + 7 + 9 + 4 + 1	42	6
2 + 6 + 3 + 2 + 3 + 6 + 2	24	6
3 + 8 + 6 + 6 + 8 + 3	34	7
4 + 1 + 9 + 1 + 4	19	1
5 + 3 + 3 + 5	16	7
6 + 5 + 6	17	8
7 + 7	14	5
8	8	8

Once again, we have patterns and symmetry. There are basically five patterns. What happens when we add our different groups of three numbers? Add these numbers: 8 + 5 + 8 = 21 and 2 + 1 = 3; 7 + 1 + 7 = 15 and 1 + 5 = 6; and then 6 + 6 + 6 = 18 and 1 + 8 = 9. So we have our favorite 3, 6, and 9 team again.

One last thing before we move on. I happened to notice something special if we add our lists of reduced diagonals together. Could we find more math magic there? Let's take a look.

Reductions +	Reductions –	Totals
1 +	8	= 9
4 +	5	= 9
1 +	8	= 9
2 +	7	= 9
8 +	1	= 9
2 +	7	= 9
3 +	6	= 9
3 +	6	= 9
3 +	6	= 9
2 +	7	= 9
8 +	1	= 9
2 +	7	= 9
1 +	8	= 9
4 +	5	= 9
1 +	8	= 9

They all add up to 9! That's pretty amazing, don't you think?

Some of these mysteries are explainable, and others are not—at least they aren't to my mind. If a pair of numbers adds up to 9, it makes sense that the corresponding multiples of those numbers would also add up to 9. Why all the diagonals add up the way they do is something I can't explain and have not taken the time to solve that mystery. I prefer to leave some mysteries unsolved. If anyone reading this can crack the code, please let me know, and we'll publish it in another book.

CHAPTER 3

Adding Consecutive Numbers

It's very beneficial to teach children strategies for adding a large group of numbers. Students who excel in math may not have difficulty with this process, but many students will struggle and get frustrated and want to give up. As many students today suffer from learning disabilities, this makes the task even more difficult.

There is an easy way to add the numbers 1 through 9, which I showed you before. Let me refresh your memory by showing you an easy way to add the numbers 1 through 8.

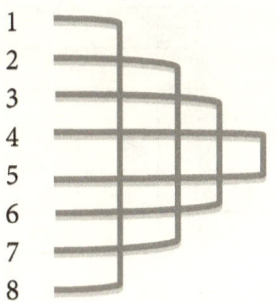

An easy way to add the numbers 1 through 8 is to make pairs that add up to 9: 1 + 8, 2 + 7, 3 + 6, and 4 + 5. That makes four 9s. Multiply 4 × 9 = 36, which reduces to 9.

All the consecutive numbers can be added this way, which greatly simplifies the process. I'm not going to show you all of these, but we will do one more—the odd number 7.

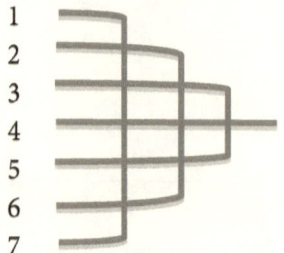

This time, we have three pairs of numbers that add up to 8: 1 + 7, 2 + 6, and 3 + 5, plus a 4 left in the middle. Multiply 3 × 8 = 24, which added to 4 equals 28.

I believe that if you add any nine consecutive numbers and then reduce those numbers, they will always reduce back to the fabulous number 9. Let's give it a try. I'm going to start at 17: 17 + 18 + 19 + 20 + 21 + 22 + 23 + 24 + 25 = 189, and then 1 + 8 + 9 = 18, and finally 1 + 8 = 9. If you still don't believe me, try it yourself. I would challenge the reader to find nine consecutive numbers that don't reduce to 9.

Let's see what happens when we add all the consecutive numbers together.

Numbers Added	Totals	Reductions
1	1	1
1+2	3	3
1+2+3	6	6
1+2+3+4	10	1
1+2+3+4+5	15	6
1+2+3+4+5+6	21	3
1+2+3+4+5+6+7	28	1
1+2+3+4+5+6+7+8	36	9
1+2+3+4+5+6+7+8+9	45	9
1+2+3+4+5+6+7+8+9+10	55	1
1+2+3+4+5+6+7+8+9+10+11	66	3
1+2+3+4+5+6+7+8+9+10+11+12	78	6
1+2+3+4+5+6+7+8+9+10+11+12+13	91	1
1+2+3+4+5+6+7+8+9+10+11+12+13+14	105	6
1+2+3+4+5+6+7+8+9+10+11+12+13+14+15	120	3
1+2+3+4+5+6+7+8+9+10+11+12+13+14+15+16	136	1
1+2+3+4+5+6+7+8+9+10+11+12+13+14+15+16+17	153	9
1+2+3+4+5+6+7+8+9+10+11+12+13+14+15+16+17+18	171	9

Can you see the repeating pattern here? We have a repeating pattern: 1, 3, 6, 1, 6, 3, 1, 9, 9. Also, if we split the 9s, there is symmetry with the numbers going up and down: 9, 1, 3, 6, 1, 6, 3, 1. I find it interesting how 3, 6, and 9 are the stars here. They definitely like to hang out together. They also invited the mighty number 1 to join the party. If we split the numbers into three groups (136, 163, and 199), you would find that each of those groups of numbers reduce to the number 1.

The numbers in the total column represent triangular numbers. Triangular numbers have that name because they always form a triangle. Technically, they are all equilateral triangles, meaning that all their sides are equal in length, even though they appear to look more like right triangles in this diagram. Here's what the first five triangular numbers look like.

Triangular Numbers

```
                                                    ×
                                ×                 × ×
                ×             × ×                 × × ×
        ×     × ×             × × ×               × × × ×
×     × ×     × × ×           × × × ×             × × × × ×
─────────────────────────────────────────────────────────
1       3       6              10                  15
```

The 3, 6, and 9 Team

The numbers 3, 6, and 9 have similar traits, and they seem to like to hang out together. The multiples of 3 and 6 when reduced both contain only the numbers 3, 6, and 9, while the multiples of 9 only reduces back to 9. All of these numbers have nice divisibility rules. If a number reduces to 9, it is divisible by 9. It will also be divisible by 3 and will also be divisible by 6 if it is an even number. A number that is even and reduces to a multiple of 3 (3, 6, or 9) will be divisible by 6 and also by 3. Numbers that reduce to a multiple of 3 (3, 6, or 9) will be divisible by 3.

On the other hand, dividing by these numbers or turning them into fractions is another story. They all turn into decimals with repeating digits. Let's look at 1/3—1 divided by 3 equals 0.33 repeating. The fraction 1/6 will reduce to 0.166, where the 6 repeats infinitely. Finally, 1/9 equals 0.11

repeating. Let's look at all the fractions that have the number 9 in the denominator.

1/9 = 0.1111
2/9 = 0.2222
3/9 = 0.3333
4/9 = 0.4444
5/9 = 0.5555
6/9 = 0.6666
7/9 = 0.7777
8/9 = 0.8888
9/9 = 1

CHAPTER 4

Perfect Squares

Perfect squares are obtained by multiplying a number by itself. A pictorial illustration of perfect squares would look like the following illustration. In each instance, the number of rows and number of columns are equal.

```
                        xxxx
                 xxx    xxxx
          xx     xxx    xxxx
x         xx     xxx    xxxx
```
1^2 2^2 3^2 4^2

Could there possibly be any magic in perfect squares? Let's take a look and see.

Number	Squared	Reduction
1	1	1
2	4	4
3	9	9
4	16	7
5	25	7
6	36	9
7	49	4
8	64	1
9	81	9

Do you remember this pattern? We've seen it before.

10	100	1
11	121	4
12	144	9
13	169	7
14	196	7
15	225	9
16	256	4
17	289	1
18	324	9

This 1, 4, 9, 7, 7, 9, 4, 1 pattern first appeared when we were looking at the positive slope diagonals.

So we have a symmetrical pattern of the first eight perfect squares. When we square 9 or a multiple of 9, the number of course reduces back to 9.

The other numbers that reduce back to 9 are, not surprisingly, the 3s and 6s and their multiples.

The pairs of numbers that add up to 9 (1 + 8, 2 + 7, 3 + 6, and 4 + 5) respectively reduce back to the same number: 1^2 and 8^2 reduce to 1; 2^2 and 7^2 reduce to 4; and 4^2 and 5^2 reduce to 7. If we add 1 + 4 + 9 + 7 + 7 + 9 + 4 + 1 = 42, they reduce to 6 (4 + 2 = 6). If we added in the extra 9 that appears for every multiple of 9, it would still reduce to 6, because 9 + 6 = 15 and 1 + 5 = 6. Once again, the number 6 is the king of the show.

There is also a pattern in subtracting consecutive perfect squares. Let's take a look at that.

Subtraction	Result	Reductions
4 - 1	3	3
9 - 4	5	5
16 - 9	7	7
25 - 16	9	9
36 - 25	11	2
49 - 36	13	4
64 - 49	15	6
81 - 64	17	8
100 - 81	19	1

The results are all consecutive odd numbers starting with the number 3. In the reductions, we're counting up by odd numbers and then counting up by even numbers. Technically, if we started the pattern with the number 1, our first number would have been the difference between 1^2 and 0^2, which would have given us 1 - 0 = 1. This pattern is similar to the pattern we got by reducing the multiples of 2.

There is another interesting pattern that is happening here when we look at the number in the results column and the numbers we are subtracting. When we have a number taken to a power, or an exponent, which in all these cases is a number squared or a number to the second power, as in 3^2, the big number in front, or the 3 in this case, is called the base, and the small number 2 is known as the exponent. There is a relationship between the base numbers and the results that we achieve by subtracting the consecutive perfect squares.

The following table will illustrate this point.

Subtraction Results	Number Squared	Base and Exponent	Addition Results
3 (4 - 1)	1	1^2	3 (1 + 2)
5 (9 - 4)	4	2^2	5 (2 + 3)
7 (16 - 9)	9	3^2	7 (3 + 4)
9 (25 - 16)	16	4^2	9 (4 + 5)
11 (36 - 25)	25	5^2	11 (5 + 6)
13 (49 - 36)	36	6^2	13 (6 + 7)
15 (64 - 49)	49	7^2	15 (7 + 8)
17 (81 - 64)	64	8^2	17 (8 + 9)
19 (100 – 81)	81	9^2	19 (9 + 10)

The "Subtraction Results" are obtained by subtracting the consecutive numbers in the "Numbers Squared" column as we saw before, and the numbers in the "Addition Results" are obtained by adding the consecutive base numbers in the "Base and Exponent" column.

Perfect Cubes

What about perfect cubes? A perfect cube is derived when we multiply a number by itself three times, as in $2 \times 2 \times 2$ for 2^3. What magic can we find there?

Number	Cubed	Reduction
1	1	1
2	8	8
3	27	9
4	64	1
5	125	8
6	216	9
7	343	1
8	512	8
9	729	9

So we have a repeating pattern of 1, 8, and 9, which will continue on past 9^3. Try it for yourself if you don't believe me.

So then I had another crazy thought. What would happen if we subtracted the differences of the cubed numbers and then reduced those results? Let's check it out.

Subtraction	Result	Reduction
8 - 1	7	7
27 - 8	19	1
64 - 27	37	1
125 - 64	61	7
216 - 125	91	1
343 - 216	127	1
512 - 343	169	7
729 - 512	217	1

So there is a repeating pattern of 7, 1, 1—more number magic. Woo-hoo! Just to make sure it continues, let's look at 10^3, which equals 1,000. If we subtract 729 from 1,000, we would get 271, which reduces to 1.

In both these instances, 9 is ruling the show since $1 + 8 + 9$ reduces to 9, and $7 + 1 + 1$ also reduces to 9. Not to mention that 7 and 11 are considered to be very lucky numbers.

CHAPTER 5

Exponents

An *exponent* is a small number that is written after a big number and tells how many times to multiply the big number by itself. It looks like this: 5^3, which you would read as five to the third power. Exponents are a more complicated concept and may be too advanced of a subject for younger learners. Please refer to my website for age-appropriate work sheets and lesson plans. Could there be any interesting patterns to be found with numbers taken to different powers? Let's explore that and see what we discover.

Nothing special is going to happen with the number 1, as it will always come out equal to 1: 1^0 or 1^{2000} will always equal 1. Also, the negative exponents for 1 will equal 1, as we are merely dividing 1 by itself. I mentioned before that there were a couple of instances where 0 has some value. Any number taken to the 0 power will equal 1. Technically, any number taken to the 0 power is like dividing a number by itself. For example: $4^0 = 1$ is like saying $4^3/4^3$ (4^3 divided by 4^3, or $64 \div 64 = 1$).

Let's move on to the number 2. Can we find any magic there?

Number w/ Exponent	Solution	Number Reduced
2^0	1	1
2^1	2	2
2^2	4	4
2^3	8	8
2^4	16	7
2^5	32	5
2^6	64	1
2^7	128	2
2^8	256	4
2^9	512	8
2^{10}	1024	7
2^{11}	2048	5

So we do find a repeating pattern of six numbers: 1, 2, 4, 8, 7, and 5.

I have grouped the numbers together to show that there are three pairs of numbers that add up to 9: 1 + 8, 2 + 7, and 4 + 5.

Adding all the numbers together: 1 + 2 + 4 + 8 + 7 + 5 = 36, which reduces to 9. Did you notice how the 3, 6, and 9 team is missing from this group of numbers?

You might be thinking, "Well, okay, but what about negative exponents? Do they follow the same pattern?" Let's take a look. In case you've forgotten what negative exponents are, it's easy to turn a negative exponent into a positive exponent by moving it to the bottom of a fraction and placing one on the top of the fraction. Here are some examples:

$2^{-1} = 1/2^1$, or just ½
$2^{-2} = 1/2^2$, or ½ × 1/2 or ¼

Number w/ Exponent	Solution	Number Reduced
2^{-6}	.015625	1
2^{-5}	.03125	2
2^{-4}	.0625	4
2^{-3}	.125	8
2^{-2}	.25	7
2^{-1}	.5	5

So, yes, we see that the same pattern of six numbers continues: 1, 2, 4, 8, 7, and 5.

Does the number 3 when taken to the different powers hold any mysteries for us? Let's explore and find out.

Number w/ Exponent	Solution	Number Reduced
3^0	1	1
3^1	3	3
3^2	9	9
3^3	27	9
3^4	81	9
3^5	243	9

Once we get to 3^3, which equals 9, each consecutive number will reduce back to 9. Taking 3^3 is like multiplying 9 × 3. As we saw with all the multiples of 9, multiplying 9 by any number will always reduce back to 9.

Now let's take a look at the negative exponents of the number 3. They are problematic because of the repeating decimal, but still there is something interesting that happens here.

Number w/ Exponent	Solution	Number Reduced
3^{-3}	.0<u>37</u>	1
3^{-2}	.11<u>1</u>	3
3^{-1}	.33<u>3</u>	9

For the first three negative exponents of the number 3, the 1, 3, 9 pattern continues, if we round to the nearest thousandth.

Let's move on to the fabulous number 4. What patterns emerge when we look at the consecutive exponents for the number 4?

Number w/ Exponent	Solution	Number Reduced
4^{-3}	.015625	1
4^{-2}	.0625	4
4^{-1}	.25	7

The negative exponents of 4 form the pattern 1, 4, 7.

4^0	1	1
4^1	4	4
4^2	16	7
4^3	64	1
4^4	256	4
4^5	1024	7
4^6	4096	1
4^7	16,384	4
4^8	65,536	7

So the exponents of four have a repeating pattern of only three numbers: 1, 4, 7. Note that $1 + 4 + 7 = 12$ and $1 + 2 = 3$. Do any of these numbers look familiar? Every exponent of 4 is the same as every other exponent of 2, as are their reduced numbers 1, 4, and 7. The number 3 plays another big part here, as going from one reduced number to the next, we add 3 each time; $1 + 3 = 4$, and then $4 + 3 = 7$, and finally $7 + 3 = 10$, which reduces back to 1.

Could there be a pattern within the exponents of the number 5? Let's find out:

Number w/ Exponent	Solution	Number Reduced
5^{-3}	.008	8
5^{-2}	.04	4
5^{-1}	.2	2

The negative exponents of 5 start a pattern of 8, 4, 2.

5^0	1	1
5^1	5	5
5^2	25	7
5^3	125	8
5^4	625	4
5^5	3,125	2
5^6	15,625	1
5^7	78,125	5
5^8	390,625	7

The pattern that was started with the negative exponents continues. The numbers that are repeating are 1, 5, 7, 8, 4, 2. These were the same numbers that made up a pattern with the exponents of the number 2, just in a slightly different order. Did you notice how the numbers 3, 6, and 9 once again are missing from this pattern?

Let's take a look at the somewhat snarky number 6. Will it reveal a pattern for us?

Number w/ Exponent	Solution	Number Reduced
6^{-3}	.004<u>629</u>	*3 or 8
6^{-2}	.02<u>7</u>	*9 or 7
6^{-1}	.1<u>6</u>	*7 or 6

* There doesn't appear to be any pattern with the negative exponents of 6. I'm showing two possible scenarios. First, 3, 9, 7 represents reducing the whole decimal, while 6, 7, 8 is just reducing the repeating part of the decimal. I believe that the repeating decimal we get with the negative exponents for the number 6 creates a problem in reducing the numbers, and therefore, no pattern is formed. They do all appear to be rational numbers, however. In case you forgot, a rational number is a number that has a terminating decimal or a repeating decimal. If there is no pattern to the decimal, then the number is irrational. The fraction 1/6 is a rational number, because it is equal to .1666 repeating. The fraction 1/17 is an irrational number, because the decimal is equal to .0588235294. There is no pattern and no repetition, and the decimal continues indefinitely.

Number w/ Exponent	Solution	Number Reduced
6^0	1	1
6^1	6	6
6^2	36	9
6^3	216	9
6^4	1,296	9
6^5	7,776	9
6^6	46,656	9

The pattern with the positive exponents of 6 is similar to what we saw with the number 3. We quickly get up to a reduced value of 9, which will continue infinitely.

Now we have one of my favorite numbers: the magic number 7. What mysteries does it have in store for us?

Number w/ Exponent	Solution	Number Reduced
7^{-3}	.0029154	2
7^{-2}	.0204081	2
7^{-1}	.<u>142857</u>	7

There doesn't appear to be a pattern to the negative exponents again. The numbers 2, 2, 7 represent a reduction through the thousandths place. Again, I believe the irrational numbers cause a problem, which interferes with a pattern being created.

7^0	1	1
7^1	7	7
7^2	49	4
7^3	343	1
7^4	2,401	7
7^5	16,807	4

We appear to have a repeating pattern with the numbers 1, 7, and 4. You might remember that these are the same three numbers that repeated in a different order for the reductions of the exponents of the number 4.

Now we have the infinitely amazing number 8. I'm sure it will have some great patterns to reveal.

Number w/ Exponent	Solution	Number Reduced
8^{-3}	.001953125	8
8^{-2}	.015625	1
8^{-1}	.125	8

The value of 8^{-3} is a little tricky, but when we add all the digits together (1 + 9 + 5 + 3 + 1 + 2 + 5), it equals 26, which reduces to 8. FYI, the decimal for 8^{-3} is a rational number.

8^0	1	1
8^1	8	8
8^2	64	1
8^3	512	8
8^4	4,096	1
8^5	32,768	8

A very simple repeating pattern is revealed with the numbers 1 and 8 repeating.

Last but definitely not least, let's take a look at the egocentric number 9. What pattern will it reveal?

Number w/ Exponent	Solution	Number Reduced
9^{-3}	.0013717421	?
9^{-2}	.012345679	1
9^{-1}	.111	1

Once again, the -3 exponent yields an irrational number. I reduced the -2 and -3 exponents by adding the repeating digits: 1 + 2 + 3 + 4 + 5 + 6 + 7 + 9 = 37/1 and just 1 for the -1 exponent.

9^0	1	1
9^1	9	9
9^2	81	9
9^3	729	9

With the positive exponents for the number 9, we quickly get up to 9 and stay there for the duration.

Here is the summary for the positive exponents of 1 through 9, including the exponent 0:

Numbers	1	2	3	4	5	6	7	8	9
	1	1	1	1	1	1	1	1	1
	1	2	3	4	5	6	7	8	9
Numbers in the Pattern	1	4	9	7	7	9	4	1	9
	1	8	9	1	8	9	1	8	9
	1	7	9	4	4	9	7	1	9
	1	5	9	7	2	9	4	8	9
Totals	6	27	40	24	27	43	24	27	46
Reductions	6	9	4	6	9	7	6	9	1

So there are some interesting things happening here. Again, the 3s, 6s, and 9s have similar patterns. I find it interesting that the 4s and 7s share the same numbers in their pattern: 1, 4, and 7. Also, the 2s and 5s share the same numbers: 1, 2, 4, 8, 7, and 5. Unlike the numbers in the multiplication table, where the pairs of numbers that added up to 9 had similar patterns, in this instance, the numbers that add up to 7 (as in 2 + 5) and 11 (4 + 7) have similar patterns. The numbers 7 and 11 have always been considered to be lucky.

Look at the pattern in the bottom line: 6, 9, 4, 6, 9, 7, 6, 9, 1. It's interesting how the 6 and 9 keep repeating, and interspersed between them is the 4, 7, and 1 from our number 4 and number 7 patterns. If we removed the reduced numbers from the 3, 6, and 9, our pattern would be 6, 9, 6, 9, 6, 9.

Here is the summary for the positive exponents of 1 through 9, excluding the exponent 0:

Numbers	1	2	3	4	5	6	7	8	9
	1	2	3	4	5	6	7	8	9
	1	4	9	7	7	9	4	1	9
Numbers in the Pattern	1	8	9	1	8	9	1	8	9
	1	7	9	4	4	9	7	1	9
	1	5	9	7	2	9	4	8	9
Totals	5	26	39	23	26	42	23	26	45
Reductions	5	8	3	5	8	6	5	8	9

Once again, an interesting pattern emerges. I find it interesting that the 3, 6, and 9 exponents reduce down to themselves, while the other numbers repeat a 5, 8 pattern. I know that many of you would like an explanation for this; however, my goal is to create an interest and excitement for your students and children. I feel that an explanation would defeat that purpose. Offer up extra credit to your students if they can come up with a logical explanation.

CHAPTER 6

The Hundreds Board

1	2	3	4	5	6	7	8	9	10
11	12	13	14	15	16	17	18	19	20
21	22	23	24	25	26	27	28	29	30
31	32	33	34	35	36	37	38	39	40
41	42	43	44	45	46	47	48	49	50
51	52	53	54	55	56	57	58	59	60
61	62	63	64	65	66	67	68	69	70
71	72	73	74	74	76	77	78	79	80
81	82	83	84	85	86	87	88	89	90
91	92	93	94	95	96	97	98	99	100

I find the hundreds board to be helpful with teaching multiplication. I like to start by having kids circle the multiples of 9 and 11. There are some nice patterns you can get from other numbers, as well, but now we're mainly interested in patterns. What mysteries does the hundreds board have in store for us? Warning: be prepared for more amazing number magic!

I took the liberty of reducing all the numbers, totaling the rows and reducing the totals of the rows, to make our work easier.

										Totals	Reductions
1	2	3	4	5	6	7	8	9	1	46	1
2	3	4	5	6	7	8	9	1	2	47	2
3	4	5	6	7	8	9	1	2	3	48	3
4	5	6	7	8	9	1	2	3	4	49	4
5	6	7	8	9	1	2	3	4	5	50	5
6	7	8	9	1	2	3	4	5	6	51	6
7	8	9	1	2	3	4	5	6	7	52	7
8	9	1	2	3	4	5	6	7	8	53	8
9	1	2	3	4	5	6	7	8	9	54	9
1	2	3	4	5	6	7	8	9	1	46	1

So we have a nice pattern with our reduced totals. We also have nice diagonal patterns, especially from lower left to upper right.

Let's look at the totals going vertically.

		1	2	3	4	5	6	7	8	9	1
		2	3	4	5	6	7	8	9	1	2
		3	4	5	6	7	8	9	1	2	3
		4	5	6	7	8	9	1	2	3	4
		5	6	7	8	9	1	2	3	4	5
		6	7	8	9	1	2	3	4	5	6
		7	8	9	1	2	3	4	5	6	7
		8	9	1	2	3	4	5	6	7	8
		9	1	2	3	4	5	6	7	8	9
		1	2	3	4	5	6	7	8	9	1
Totals		46	47	48	49	50	51	52	53	54	46
Reductions		1	2	3	4	5	6	7	8	9	1

Adding the columns gives us the same pattern as adding the rows. This is not surprising, as we are adding the same numbers in each instance.

So what would happen if we added up the diagonals from lower left to upper right or in a positive slope direction? Let's give it a shot. Once again, there are some interesting patterns that have formed, with the same number being repeated diagonally. We'll start off with one 1, and then two 2s, and then 3 threes, and so on up to 9.

1	2	3	4	5	6	7	8	9	1
2	3	4	5	6	7	8	9	1	2
3	4	5	6	7	8	9	1	2	3
4	5	6	7	8	9	1	2	3	4
5	6	7	8	9	1	2	3	4	5
6	7	8	9	1	2	3	4	5	6
7	8	9	1	2	3	4	5	6	7
8	9	1	2	3	4	5	6	7	8
9	1	2	3	4	5	6	7	8	9
1	2	3	4	5	6	7	8	9	1

Let's see if this pattern of numbers gives us another pattern in our reductions.

Number	Quantity	Multiplication	Total	Reduction
1	1	1 × 1	1	1
2	2	2 × 2	4	4
3	3	3 × 3	9	9
4	4	4 × 4	16	7
5	5	5 × 5	25	7
6	6	6 × 6	36	9
7	7	7 × 7	49	4
8	8	8 × 8	64	1
9	9	9 × 9	81	9
1	10	1 × 10	10	1
2	9	2 × 9	18	9
3	8	3 × 8	24	6
4	7	4 × 7	28	1
5	6	5 × 6	30	3
6	5	6 × 5	30	3
7	4	7 × 4	28	1
8	3	8 × 3	24	6
9	2	9 × 2	18	9
1	1	1 × 1	1	1

There appears to be two different symmetrical patterns that develop in our reduced numbers. I might note that we've seen both of these patterns before. The pattern 1, 4, 9, 7, 7, 9, 4, 1 appeared when we were working with perfect squares, which makes sense, as those numbers also represent perfect squares (1 × 1, 2 × 2, 3 × 3, etc.). The pattern 6, 1, 3, 3, 1, 6 was revealed in two of the diagonals of our reduced multiples.

So let's take a look at the other diagonals. If we start at the top right-hand corner, we have 1 and then 9, 2. If you look to the lower left-hand corner, you will find the same numbers. Let's add the diagonals together and see if we can find another pattern.

1	2	3	4	5	6	7	8	9	1
2	3	4	5	6	7	8	9	1	2
3	4	5	6	7	8	9	1	2	3
4	5	6	7	8	9	1	2	3	4
5	6	7	8	9	1	2	3	4	5
6	7	8	9	1	2	3	4	5	6
7	8	9	1	2	3	4	5	6	7
8	9	1	2	3	4	5	6	7	8
9	1	2	3	4	5	6	7	8	9
1	2	3	4	5	6	7	8	9	1

This table shows the diagonal numbers added together and reduced in a negative slope direction.

Numbers	Total	Reduction
1	1	1
9 + 2	11	2
8 + 1 + 3	12	3
7 + 9 + 2 + 4	22	4
6 + 8 + 1 + 3 + 5	23	5
5 + 7 + 9 + 2 + 4 + 6	33	6
4 + 6 + 8 + 1 + 3 + 5 + 7	34	7
3 + 5 + 7 + 9 + 2 + 4 + 6 + 8	44	8
2 + 4 + 6 + 8 + 1 + 3 + 5 + 7 + 9	45	9
1 + 3 + 5 + 7 + 9 + 2 + 4 + 6 + 8 + 1	46	1
2 + 4 + 6 + 8 + 1 + 3 + 5 + 7 + 9	45	9
3 + 5 + 7 + 9 + 2 + 4 + 6 + 8	44	8
4 + 6 + 8 + 1 + 3 + 5 + 7	34	7
5 + 7 + 9 + 2 + 4 + 6	33	6
6 + 8 + 1 + 3 + 5	23	5
7 + 9 + 2 + 4	22	4
8 + 1 + 3	12	3
9 + 2	11	2
1	1	1

Once again, we have all kinds of patterns and symmetry and magic and not just in the final reduction but in every column. Notice the numbers vertically, horizontally, and along the edges in the first column. In the totals column, see how first the ones' digit is repeated and then the tens' digit is repeated. This gets interrupted in the middle with the total of 46 and then starts up again symmetrical to the top half of the column. The reduction column has a nice symmetrical pattern counting from 1 up to 9, ending with a 1 in the middle and then counting back down to 1. Pretty amazing, I'd say.

In comparing the two diagonals of reduced numbers, I don't find a pattern with adding them together, but I do find that they have the same numbers in the beginning, the middle, and the end, which have been highlighted below.

1	1
2	4
3	9
4	7
5	7
6	9
7	4
8	1
9	9
1	1
9	9
8	6
7	1
6	3
5	3
4	1
3	6
2	9
1	1

I do find a relationship between these numbers. Do you know what it is?

Give it a try before I reveal what I found.

2	4
3	9
4	7
5	7
6	9
7	4
8	1

Original Number	Squared	Reduced	Final Number
2	4	4	4
3	9	9	9
4	16	1 + 6	7
5	25	2 + 5	7
6	36	3 + 6	9
7	49	4 + 9 / 1 + 3	4
8	64	6 + 4 / 1 + 0	1

This reveals more amazing math. As you can see, the relationship between the numbers is that the second number equals the first number squared and then reduced. Who knew there were so many hidden patterns in numbers?

I can find a vertical relationship between the columns of numbers. See if your students can find them. The first column contains these numbers: 1 2 3 4 5 6 7 8 9 1 9 8 7 6 5 4 3 2 1. The second column contains these numbers: 1 4 9 7 7 9 4 1 9 6 1 3 3 1 6 9 1. If you can't solve it, refer to my website for a video on the solutions.

This final table of numbers gave me a little brain workout. I definitely couldn't find a solution using my normal techniques. There is a very different method required to solve this pattern. Let me know if you figure it out. You can also visit my website for a video showing the solution.

8	6
7	1
6	3
5	3
4	1
3	6
2	9

CONCLUSIONS

I knew there were some interesting patterns to be found in the multiples of numbers; however, I didn't realize how extensive they were until I started to write them all down. I also didn't realize there were patterns in perfect squares, perfect cubes, and exponents. I'll let you draw your own conclusions about all the patterns that have emerged. Personally, I don't think these are random occurrences. Mathematically, it isn't possible. The probability of all these patterns is off the chart.

I'm sure many of these patterns can be explained and formulas could be derived, but for me, that would ruin this process. As I've told you all along, I want to make math fun. I want kids to have a reason to want to dig deeper, to explore the mysteries. If your children have more fun with math, I have succeeded in what I set out to do.

I was aware of numerology before I wrote this book; it is not a well-known or well-respected science. However, through the process of writing this book, I have gained respect for numerology. Math is normally taught as a pretty cut-and-dried subject. I recommend that you play with it, have fun with it, and don't try to explain it all. Let the young, creative minds have fun and enjoy the learning process.

WORKSHEETS

The following worksheets are designed to be used with students primarily in grades 3 through 6. They can also be used with struggling students in higher grades. I have found students to really enjoy working with these types of problems. A few of the worksheets can also be used with younger students. The best worksheets for younger students would be the first 2 and the coloring worksheets: 13 – 15. The last 2 worksheets on exponents may be a challenge for grades 3 and 4, but these students can do the work with a little instruction and encouragement. Telling your students that this is work normally reserved for older students will bring an air of excitement and challenge for your students. All of the worksheets are similar to problems shown in the book.

I hope these help your struggling students to become more engaged and excited with their math. Please send your comments and testimonials to my website: www.mindblowingmath.com Remember to visit the site also for blogs and videos with more exciting work.

Name_____   Date_____

Adding the numbers 1 thru 9   Worksheet #1

Draw lines to connect the numbers that add up to 10.

1

2

3

4

5

6

7

8

9

List the pairs of numbers that add up to 10.

_____ + _____ = 10

_____ + _____ = 10

_____ + _____ = 10

_____ + _____ = 10

Circle any number that did not connect to another number to add up to 10.

What do all the numbers add up to? _____

Name_____   Date_____

Reducing Numbers   **Worksheet #2**

All numbers reduce down to a single digit. This is done by adding the digits
of a number together, until you just have one number. Usually you only
have to do this once, but there are times when you will have to reduce twice.

Examples:
Reducing once:
10 . . . .1 + 0 = 1   22. . . . 2 + 2 = 4
11 . . . .1 + 1 = 2   27. . . . 2 + 7 = 9

Reducing twice
38. . . . 3 + 8 = 11, 1+1 = 2
49. . . . 4 + 9 = 13, 1+ 3 = 4

Reduce these numbers down to a single digit. (Pretend like there's a plus
sign between the two numbers)

54_____   33_____   61_____

72_____   66_____   48_____

34_____   29_____   43_____

87_____   91_____   17_____

12_____   46_____   41_____

59_____   25_____   53_____

67_____   99_____   87_____

24_____   23_____   51_____

77_____   69_____   47_____

93_____   16_____   63_____

Name_____   Date_____

The Multiples of 2   **Worksheet #3**

Solve the multiplication
problems.

Reduce and rewrite your
answers as a single digit.

2 x 1 = _____   2 x 1 = _____

2 x 2 = _____   2 x 2 = _____

2 x 3 = _____   2 x 3 = _____

2 x 4 = _____   2 x 4 = _____

2 x 5 = _____   2 x 5 = _____

2 x 6 = _____   2 x 6 = _____

2 x 7 = _____   2 x 7 = _____

2 x 8 = _____   2 x 8 = _____

2 x 9 = _____   2 x 9 = _____

List your answers in order from the 2nd column.

_____

Is there a pattern? _____

Describe the pattern.

_____

_____

_____

Name_____   Date_____

The Multiples of 3   **Worksheet #4**

Solve the multiplication   Reduce and rewrite your
problems.   answers as a single digit.

3 x 1 = _____   3 x 1 = _____

3 x 2 = _____   3 x 2 = _____

3 x 3 = _____   3 x 3 = _____

3 x 4 = _____   3 x 4 = _____

3 x 5 = _____   3 x 5 = _____

3 x 6 = _____   3 x 6 = _____

3 x 7 = _____   3 x 7 = _____

3 x 8 = _____   3 x 8 = _____

3 x 9 = _____   3 x 9 = _____

List your answers in order from the 2nd column.

_____

Is there a pattern? _____

Describe the pattern.

_____

_____

_____

Name_____   Date_____

The Multiples of 4   **Worksheet #5**

Solve the multiplication
problems.

Reduce and rewrite your
answers as a single digit.

4 x 1 = _____   4 x 1 = _____

4 x 2 = _____   4 x 2 = _____

4 x 3 = _____   4 x 3 = _____

4 x 4 = _____   4 x 4 = _____

4 x 5 = _____   4 x 5 = _____

4 x 6 = _____   4 x 6 = _____

4 x 7 = _____   4 x 7 = _____

4 x 8 = _____   4 x 8 = _____

4 x 9 = _____   4 x 9 = _____

List your answers in order from the 2nd column.

_____

Is there a pattern? _____

Describe the pattern.

_____

_____

_____

Name_____   Date_____

The Multiples of 5   ## Worksheet #6

Solve the multiplication problems.

Reduce and rewrite your answers as a single digit.

5 x 1 = _____   5 x 1 = _____

5 x 2 = _____   5 x 2 = _____

5 x 3 = _____   5 x 3 = _____

5 x 4 = _____   5 x 4 = _____

5 x 5 = _____   5 x 5 = _____

5 x 6 = _____   5 x 6 = _____

5 x 7 = _____   5 x 7 = _____

5 x 8 = _____   5 x 8 = _____

5 x 9 = _____   5 x 9 = _____

List your answers in order from the 2nd column.

_____

Is there a pattern? _____

Describe the pattern.

_____

_____

_____

Name_____   Date_____

The Multiples of 6   **Worksheet #7**

Solve the multiplication
problems.

Reduce and rewrite your
answers as a single digit.

6 x 1 = _____

6 x 2 = _____

6 x 3 = _____

6 x 4 = _____

6 x 5 = _____

6 x 6 = _____

6 x 7 = _____

6 x 8 = _____

6 x 9 = _____

6 x 1 = _____

6 x 2 = _____

6 x 3 = _____

6 x 4 = _____

6 x 5 = _____

6 x 6 = _____

6 x 7 = _____

6 x 8 = _____

6 x 9 = _____

List your answers in order from the 2nd column.

_____

Is there a pattern? _____

Describe the pattern.

_____

_____

Is this pattern similar to another pattern we found? _____

Name_____   Date_____

The Multiples of 7   ## Worksheet #8

Solve the multiplication   Reduce and rewrite your
problems.   answers as a single digit.

7 x 1 = _____   7 x 1 = _____

7 x 2 = _____   7 x 2 = _____

7 x 3 = _____   7 x 3 = _____

7 x 4 = _____   7 x 4 = _____

7 x 5 = _____   7 x 5 = _____

7 x 6 = _____   7 x 6 = _____

7 x 7 = _____   7 x 7 = _____

7 x 8 = _____   7 x 8 = _____

7 x 9 = _____   7 x 9 = _____

List your answers in order from the 2nd column.

_____

Is there a pattern? _____

Describe the pattern.

_____

_____

_____

Name_____   Date_____

The Multiples of 8   ## Worksheet #9

Solve the multiplication problems.

Reduce and rewrite your answers as a single digit.

8 x 1 = _____   8 x 1 = _____

8 x 2 = _____   8 x 2 = _____

8 x 3 = _____   8 x 3 = _____

8 x 4 = _____   8 x 4 = _____

8 x 5 = _____   8 x 5 = _____

8 x 6 = _____   8 x 6 = _____

8 x 7 = _____   8 x 7 = _____

8 x 8 = _____   8 x 8 = _____

8 x 9 = _____   8 x 9 = _____

List your answers in order from the 2nd column.

_____

Is there a pattern? _____

Describe the pattern.

_____

_____

_____

Name_____   Date_____

The Multiples of 9   **Worksheet #10**

Solve the multiplication   Reduce and rewrite your
problems.   answers as a single digit.

9 x 1 = _____   9 x 1 = _____

9 x 2 = _____   9 x 2 = _____

9 x 3 = _____   9 x 3 = _____

9 x 4 = _____   9 x 4 = _____

9 x 5 = _____   9 x 5 = _____

9 x 6 = _____   9 x 6 = _____

9 x 7 = _____   9 x 7 = _____

9 x 8 = _____   9 x 8 = _____

9 x 9 = _____   9 x 9 = _____

List your answers in order from the 2nd column.

_____

Is there a pattern? _____

Describe the pattern. _____

What patterns do you see with your answers in the first column?

_____

_____

Name_____   Date_____

Multiplication Table   **Worksheet #11**

Fill in the blanks in the multiplication table below.

×	1	2	3	4	5	6	7	8	9	10
1										
2										
3										
4										
5										
6										
7										
8										
9										
10										

Name_____   Date_____

Multiplication Table Reduced   **Worksheet #12**

Fill in the blanks in the multiplication table below, using the reduced number solutions. (Note: I eliminated the column and row for 9 to make the patterns more visible.)

×	1	2	3	4	5	6	7	8
1								
2								
3								
4								
5								
6								
7								
8								

What patterns do you see in this table?

_____

_____

_____

Name_____   Date_____

Coloring Patterns   **Worksheet #13**

Color in the table below. Be sure to use the same color for individual
numbers. For example: color all 1s red. Fill in the key with the same colors.

1	2	3
4	5	6
7	8	9

1	2	3	4	5	6	7	8
2	4	6	8	1	3	5	7
3	6	9	3	6	9	3	6
4	8	3	7	2	6	1	5
5	1	6	2	7	3	8	4
6	3	9	6	3	9	6	3
7	5	3	1	8	6	4	2
8	7	6	5	4	3	2	1

Name_____   Date_____

Coloring Patterns 2   **Worksheet #14**

Choose four colors to fill in the table below. Be sure to color your key as well.

1, 8	
2, 7	
3, 6	
4, 5	

1	2	3	4	5	6	7	8
2	4	6	8	1	3	5	7
3	6	9	3	6	9	3	6
4	8	3	7	2	6	1	5
5	1	6	2	7	3	8	4
6	3	9	6	3	9	6	3
7	5	3	1	8	6	4	2
8	7	6	5	4	3	2	1

Name_____   Date_____

Coloring Patterns 3   **Worksheet #15**

Choose three colors to fill in the table below. Be sure to color your key as well.

1, 4, 7	
2, 5, 8	
3, 6, 9	

1	2	3	4	5	6	7	8
2	4	6	8	1	3	5	7
3	6	9	3	6	9	3	6
4	8	3	7	2	6	1	5
5	1	6	2	7	3	8	4
6	3	9	6	3	9	6	3
7	5	3	1	8	6	4	2
8	7	6	5	4	3	2	1

Name_____   Date_____

Perfect Squares   **Worksheet #16**

Fill in the following table by first solving for the perfect square and then reduce those perfect squares down to a single digit number.

Number	Number Squared	Square Reduced
1	1 x 1 =	
2	2 x 2 =	
3	3 x 3 =	
4	4 x 4 =	
5	5 x 5 =	
6	6 x 6 =	
7	7 x 7 =	
8	8 x 8 =	

Is there a pattern?_____

Describe the pattern_____

_____

Name_____   Date_____

Perfect Cubes   **Worksheet #17**

Fill in the following table by first solving for the perfect square and then reduce those perfect squares down to a single digit number.

Number	Number Cubed	Cube Reduced
1	1 x 1 x 1 =	
2	2 x 2 x 2 =	
3	3 x 3 x 3 =	
4	4 x 4 x 4 =	
5	5 x 5 x 5 =	
6	6 x 6 x 6 =	
7	7 x 7 x 7 =	
8	8 x 8 x 8 =	
9	9 x 9 x 9 =	

Is there a pattern?_____

Describe the pattern_____

_____

Name_____   Date_____

Adding groups of numbers   **Worksheet #18**

Add each row of numbers and put the answer in the "Totals" column. Use grouping techniques to make the job easier. Then reduce the numbers in the totals column to a single digit number.

Numbers	Totals	Reductions
1		
2 + 2		
3 + 4 + 3		
4 + 6 + 6 + 4		
5 + 8 + 9 + 8 + 5		
6 + 1 + 3 + 3 + 1 + 6		
7 + 3 + 6 + 7 + 6 + 3 + 7		
8 + 5 + 9 + 2 + 2 + 9 + 5 + 8		
7 + 3 + 6 + 7 + 6 + 3 + 7		
6 + 1 + 3 + 3 + 1 + 6		
5 + 8 + 9 + 8 + 5		
4 + 6 + 6 + 4		
3 + 4 + 3		
2 + 2		
1		

What patterns are revealed in the Totals and the Reductions columns?

_____

_____

_____

Name_____   Date_____

Exponents of the Number 2   Worksheet #19

Solve for the exponents, then reduce your solutions.

Number with Exponent	Solution	Solution Reduced
2^0 (2/2)		
2^1 (2)		
2^2 (2 x 2)		
2^3 (2 x 2 x 2)		
2^4 (2 x 2 x 2 x 2)		
2^5 (2 x 2 x 2 x 2 x 2)		

Are you up for a challenge? Complete the next table.

Number with Exponent	Solution	Solution Reduced
2^6 (2 x 2 x 2 x 2 x 2 x 2)		
2^7 (2 x 2 x 2 x 2 x 2 x 2 x 2)		
2^8 (2 x 2 x 2 x 2 x 2 x 2 x 2 x 2)		
2^9 (2 x 2 x 2 x 2 x 2 x 2 x 2 x 2 x 2)		
2^{10} (2 x 2 x 2 x 2 x 2 x 2 x 2 x 2 x 2 x 2)		
2^{11} (2 x 2 x 2 x 2 x 2 x 2 x 2 x 2 x 2 x 2 x 2)		

Is there a pattern in your reduced solutions? _____

Describe the pattern? _____

_____

What numbers are missing from this pattern? _____

Name_____   Date_____

Exponents of the Number 3   **Worksheet #20**

Solve for the exponents, then reduce your solutions.

Number with Exponent	Solution	Solution Reduced
3^0 (3/3)		
3^1 (3)		
3^2 (3 x 3)		
3^3 (3 x 3 x 3)		
3^4 (3 x 3 x 3 x 3)		
3^5 (3 x 3 x 3 x 3 x 3)		
3^6 (3 x 3 x 3 x 3 x 3 x 3)		

Are your solutions even, odd or both even and odd? _____

Is there a pattern to your solutions?_____

Why do you think so many of your solutions are the same? _____

_____

_____

www.ingramcontent.com/pod-product-compliance
Lightning Source LLC
Chambersburg PA
CBHW030855180526
45163CB00004B/1578